Addition (HTU, odd/even)

Complete these patterns.

1 odd numbers 1 3 ☐ 7 ☐ 11 ☐ 15 ☐ 19

2 even numbers 2 4 ☐ 8 10 ☐ 14 ☐ ☐ 20

3 The sixth odd number is ☐

4 The tenth odd number is ☐

5 The second even number is ☐

6 The seventh even number is ☐

7 Write 4 pairs of odd numbers which add up to 16.

☐ + ☐   ☐ + ☐   ☐ + ☐   ☐ + ☐

8 Write 3 pairs of even numbers which add up to 16.

☐ + ☐   ☐ + ☐   ☐ + ☐

9 Write an odd and an even number which add up to 15.

☐ + ☐ = 15
odd   even

10 Find other ways to do it.

Add these pairs of numbers.

11
```
H T U
1 9 0
+ 2 5 1
———————
```

12
```
H T U
2 8 1
+ 1 6 1
———————
```

13
```
H T U
1 5 0
+ 2 9 5
———————
```

14 ☐ 272 + ☐ 171 = ☐

15 ☐ 162 + ☐ 282 = ☐

16 Which answers are odd numbers?
Write them in order.

Harder addition

Write which of these are odd numbers and which are even numbers.

1 241 _____

2 572 _____

3 130 _____

4 174 _____

5 935 _____

6 489 _____

Use the numbers 1, 2, 3, 4 once each time.
Do the sums.

```
   T  U
   1  2   even
+  4  3   odd
   5  5   odd
```

7
```
   T   U
 [ ] [ ]  odd
+[ ] [ ]  odd
 _____
```

8
```
   T   U
 [ ] [ ]  even
+[ ] [ ]  even
 _____
```

9
```
  H  T  U
  2  5  3
+ 1  6  4
_____
```

10
```
  H  T  U
  1  3  5
+ 1  7  0
_____
```

11
```
  H  T  U
  2  5  6
+ 1  8  3
_____
```

12
```
  H  T  U
  1  7  1
+ 2  6  3
_____
```

13
```
  H  T  U
  2  8  5
+ 1  4  1
_____
```

14
```
  H  T  U
  2  9  3
+ 1  3  4
_____
```

15
```
  H  T  U
  2  1  2
+ 1  9  3
_____
```

16
```
  H  T  U
  1  4  7
+ 2  6  1
_____
```

17
```
  H  T  U
  2  9  7
+ 1  6  2
_____
```

18 How many answers are odd numbers? [ ]

19 How many answers are even numbers? [ ]

20 Make up a sum. It must have an even answer.

Subtraction (HTU, patterns)

1 The difference between 7 and 3 is ☐

Find the difference between these numbers.

2 6 and 9 ☐

3 16 and 20 ☐

4 27 and 33 ☐

5 46 and 53 ☐

6 There are 17 children.
9 are in one team. How many are in the other? ☐

7 There are 253 children.
126 are boys. How many are girls? ☐

1	2	3	4	5	6	7	8	9	10
11	12	13	14	15	16	17	18	19	20
21	22	23	24	25	26	27	28	29	30

8 Count back in 3s. Start at 29.

29 26 ☐ ☐ ☐ ☐ ☐ ☐ ☐ ☐

9 Count back in 4s. Start at 27.

27 23 ☐ ☐ ☐ ☐ ☐

10 Count back in 7s. Start at 30.

30 23 ☐ ☐ ☐

11
```
  T U
  9 6
- 3 7
-----
```

12
```
  T U
  4 2
- 1 9
-----
```

13
```
  T U
  6 3
- 2 8
-----
```

14
```
  T U
  4 5
- 1 7
-----
```

15
```
  H T U
  2 4 8
-   2 9
-------
```

16
```
  H T U
  2 8 5
- 1 4 7
-------
```

17
```
  H T U
  3 7 5
- 1 5 8
-------
```

Harder subtraction

1 John has 125 sweets and gives away 7. ☐
How many are left?

2 Ann has 132 sweets and gives away 18. ☐
How many are left?

What is the difference between the
scores of the teams?

Points	
orange team	192
green team	168
blue team	149
yellow team	93
red team	88

3 yellow and red ☐

4 green and blue ☐

5 orange and green ☐

6 orange and blue ☐

7 orange and red ☐

Do these in your head. Check with a calculator.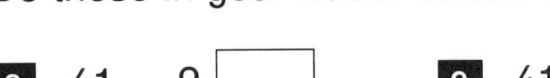

8 41 − 9 ☐

9 41 − 19 ☐

10 41 − 29 ☐

11 53 − 9 ☐

12 53 − 19 ☐

13 53 − 29 ☐

14 60 − 9 ☐

15 60 − 19 ☐

16 60 − 29 ☐

17
```
  H  T  U
  3  4  2
- 1  2  5
_____
```

18
```
  H  T  U
  2  6  6
- 1  3  8
_____
```

19
```
  H  T  U
  3  8  4
- 1  4  7
_____
```

20
```
  H  T  U
  3  8  3
- 1  2  9
_____
```

21
```
  H  T  U
  4  2  5
- 2  1  6
_____
```

22
```
  H  T  U
  5  4  0
- 1  2  9
_____
```

Multiplication (patterns for 6, 9)

1   1 box has ☐ eggs

      2 boxes have ☐ eggs

      3 boxes have ☐ eggs

2   Colour the pattern of 6.

1	2	3	4	5	6
7	8	9	10	11	12
13	14	15	16	17	18
19	20	21	22	23	24
25	26	27	28	29	30
31	32	33	34	35	36
37	38	39	40	41	42
43	44	45	46	47	48
49	50	51	52	53	54
55	56	57	58	59	60

3 $4 \times 6 =$ ☐   **4** $6 \times 6 =$ ☐

5 $9 \times 6 =$ ☐   **6** $7 \times 6 =$ ☐

7 $8 \times 6 =$ ☐   **8** $10 \times 6 =$ ☐

9 $5 \times 6 =$ ☐   **10** $3 \cdot \times 6 =$ ☐

11
```
H  T  U
   2  2
×        6
_____
```

12
```
H  T  U
   2  5
×        6
_____
```

13
```
H  T  U
   2  3
×        6
_____
```

14
```
H  T  U
   3  1
×        6
_____
```

15 $9 + 9 =$ ☐

      $2 \times 9 =$ ☐

16 $9 + 9 + 9 =$ ☐

      ☐ $\times 9 =$ ☐

17 $9 + 9 + 9 + 9 =$ ☐

      ☐ $\times 9 =$ ☐

18 $9 + 9 + 9 + 9 + 9 =$ ☐

      ☐ $\times 9 =$ ☐

19
```
H  T  U
   2  1
×        9
_____
```

20
```
H  T  U
   3  2
×        9
_____
```

21
```
H  T  U
   4  3
×        9
_____
```

22
```
H  T  U
   2  5
×        9
_____
```

Harder multiplication

1 Join the numbers in the pattern of 6.

6	18	18	36	45	54	42	48	81	60
9	12	27	24	30	36	63	72	54	90

2 Join the numbers in the pattern of 9.

3 Which three numbers are in the patterns of 6 and 9?

Find the quickest way to do these.

4 $9 + 9 + 9 + 9 + 9 + 9$ $\boxed{} \times 9 = \boxed{}$

5 $6 + 6 + 6 + 6 + 6 + 6 + 6 + 6 + 6$ $\boxed{} \times 6 = \boxed{}$

6 What do you notice about the answers?

7 $37 + 37 + 37 + 37 + 37 + 37$ $\boxed{} \times 37 = \boxed{}$

8 $29 + 29 + 29 + 29 + 29 + 29$ $\boxed{} \times \boxed{} = \boxed{}$

9 $25 + 25 + 25 + 25 + 25 + 25 + 25 + 25 + 25$

$\boxed{} \times \boxed{} = \boxed{}$

10
```
  H  T  U
     2  1
×        6
─────────
```

11
```
  H  T  U
     1  4
×        9
─────────
```

12
```
  H  T  U
     3  9
×        6
─────────
```

13
```
  H  T  U
     3  0
×        9
─────────
```

14
```
  H  T  U
     4  5
×        6
─────────
```

15
```
  H  T  U
     5  3
×        9
─────────
```

16 How many pairs of answers are the same? $\boxed{}$

Division (patterns for 6, 9, TU by 1 digit)

1 How many counters are there? ☐

2 How many sixes are there? ☐

Complete the patterns.

3 6   12   18 ☐ ☐ ☐ 42 ☐ 54 ☐

4 9   18   27 ☐ ☐ 54 63 ☐ 81 ☐

5 $30 \div 6 =$ ☐   **6** $54 \div 6 =$ ☐   **7** $36 \div 6 =$ ☐

8 $81 \div 9 =$ ☐   **9** $36 \div 9 =$ ☐   **10** $72 \div 9 =$ ☐

11 $\dfrac{24}{6} =$ ☐   **12** $\dfrac{60}{6} =$ ☐   **13** $\dfrac{27}{9} =$ ☐   **14** $\dfrac{45}{9} =$ ☐

15 $2\overline{)26}$ **16** $2\overline{)42}$ **17** $3\overline{)36}$ **18** $3\overline{)66}$

Harder

1 $18 \div 2 =$ ☐ $9 \times$ ☐ $= 18$

$18 \div 9 =$ ☐ $2 \times$ ☐ $= 18$

2 Put 24 children into groups of 6. How many groups? ☐

3 Put 24 children into groups of 4. How many groups? ☐

4 $2\overline{)32}$ **5** $2\overline{)38}$ **6** $2\overline{)56}$ **7** $3\overline{)42}$

8 $3\overline{)45}$ **9** $3\overline{)54}$ **10** $4\overline{)52}$ **11** $4\overline{)64}$

Money (addition/subtraction of pounds/pence)

Add up these bills.

1 £
  1 · 2 5
  1 · 4 9
_____

2 £
  2 · 3 6
  3 · 2 7
_____

3 £
  1 · 5 8
  3 · 1 2
_____

4 £
  2 · 7 5
  2 · 5 0
_____

5 £
  4 · 6 2
  3 · 2 9
_____

6 £
  5 · 3 9
  2 · 1 4
_____

How many coins are in these bags?

7 [ ]   **8** [ ]   **9** [ ]   **10** [ ]

50p
in
2p coins

80p
in
10p coins

£1·20p
in
20p coins

£1·50p
in
50p coins

11 Mary has £1·72. What coins can she have?
Draw them.

Here are some bills.
What is the change each time from £1·20?

12
Pizza

£1·12

13
Potato

95p

14
Ice cream

89p

Harder money

How much are these bills?

1 £
1 · 2 3
1 · 4 5
2 · 1 6
—————

2 £
1 · 1 2
2 · 4 3
1 · 2 7
—————

3 £
2 · 3 8
1 · 1 4
2 · 2 1
—————

4 George has £1·55.
He buys one drink and
has 20p left. Which
drink did he buy?

5 Jean has £1·10. She buys
a small drink and has
two coins left. Which are they?

> **Drinks**
> large £1·35
> small £0·95

6

	£
Ali has	1 · 6 0
A drink costs	1 · 3 5
The money left is	————

7

	£
Saeed has	1 · 9 9
A drink costs	0 · 9 5
The money left is	————

8 How much do the coins add up to?   £0 · ☐

Use four different coins each time.

9 What different amounts can you make?

Weight (addition/subtraction of grams)

Find the weights.

1	g	**2**	g	**3**	g
apple	1 2 5	mango	1 4 4	banana	1 2 6
crisps	2 8	chocolate	6 5	pear	1 1 9
Total weight	____	Total weight	____	Total weight	____

4 g
```
   2 3 8
 + 4 5 7
 -------
```

5 g
```
   5 3 7
 + 1 2 6
 -------
```

6 g
```
   4 4 2
 + 1 7 3
 -------
```

7 g
```
   3 3 3
 + 3 8 4
 -------
```

8 g
```
   4 5 6
 - 1 2 8
 -------
```

9 g
```
   3 2 8
 - 1 1 9
 -------
```

10 g
```
   4 6 0
 - 1 3 2
 -------
```

11 g
```
   5 3 4
 - 2 0 5
 -------
```

Harder

Add the weights each bought.

1  Jo

cheese	2 1 8 g
fruit	2 5 0 g
pasta	2 1 2 g

2  Viv

beans	4 2 0 g
sauce	3 4 0 g
bread	2 1 3 g

3  Tom

soup	2 9 5 g
biscuits	1 5 0 g
cake	1 8 0 g

4 Who bought 680g? _____

5 Who has the heaviest bag? _____

6 Who has the lightest bag? _____

7 How much lighter is Tom's bag than Viv's bag? ☐ g

8 How much heavier is Jo's bag than Tom's bag? ☐ g

Time (5-minute intervals)

Show the times.

1 5 past 10

2 $\frac{1}{2}$ past 3

3 $\frac{1}{4}$ past 7

4 25 past 4

5 Draw arrows to match the times.

8:20   7:30   5:15   11:00

Write these times.

6

7

8

9
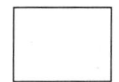

[ ] to [ ]   [ ] to [ ]   [ ] to [ ]   [ ] to [ ]

Show these times on the clock faces.

10 8:40

11 4:45
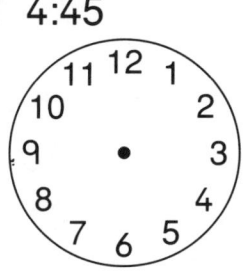

12 11:25
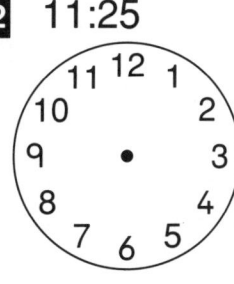

13 8:50
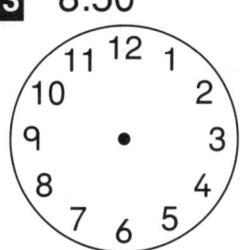

Harder time

Fill in the missing times.

1

| 9:00 | o'clock |

2

| | 25 to 6 |

3

| 3:35 | |

4

| | |

5

| | quarter to 4 |

6

| 10:55 | |

7

| | 5 to 8 |

Addition (HTU, place value, patterns)

1
```
  H  T  U
  2  3  4
+ 1  4  7
─────────
```

2
```
  H  T  U
  5  5  6
+ 2  2  5
─────────
```

3
```
  H  T  U
  3  4  6
+ 4  9  2
─────────
```

4
```
  H  T  U
  6  7  1
+ 1  7  4
─────────
```

5
```
  H  T  U
  1  8  5
+ 2  3  6
─────────
```

6
```
  H  T  U
  7  1  3
+ 1  9  8
─────────
```

7
```
  H  T  U
  4  7  1
+ 2  5  9
─────────
```

8
```
  H  T  U
  6  5  8
+ 1  6  6
─────────
```

9 346 + 275

```
 H  T  U


+ _____
  _____
```

10 394 + 137

```
 H  T  U


+ _____
  _____
```

11 562 + 189

```
 H  T  U


+ _____
  _____
```

→ 200 + 40 + 3 = 243

12

→ ☐ + ☐ + ☐ = ☐

13

→ ☐ + ☐ + ☐ = ☐

14

→ ☐ + ☐ + ☐ = ☐

Complete the patterns.

15 Add 5.   4   9   ☐ ☐ ☐ ☐ ☐ ☐

16 Add 10.   9   ☐ ☐ ☐ ☐ ☐ ☐ ☐

Harder addition

| 300 | 60 | 9 | → | 369 |

1 | 200 | 40 | 3 | → | [ ] |

2 | [ ] | 30 | 4 | → | 434 |

3 | 500 | [ ] | [ ] | → | 567 |

4 | [ ] | 50 | [ ] | → | 653 |

5 two + three hundred + forty = [ ]

6 four hundred + one + twenty = [ ]

7 thirty + six hundred + five = [ ]

8
```
  H  T  U
  5  3  2
+ 2  7  7
---------
```

9
```
  H  T  U
  5  0  3
+ 1  9  7
---------
```

10
```
  H  T  U
  4  8  7
+ 3  3  3
---------
```

11
```
  H  T  U
  6  1  4
+ 2  8  8
---------
```

12
```
  H  T  U
  4  5  5
+ 4  5  5
---------
```

13
```
  H  T  U
  3  9  7
+ 1  5  4
---------
```

14
```
  H  T  U
  3  7  7
+ 3  3  3
---------
```

15
```
  H  T  U
  1  3  6
+ 1  7  5
---------
```

```
        +
| 3  4  7 | 2  8  4 |
| 5  0  6 | 2  9  5 |
| 7  7  5 | 1  5  5 |
```

16
```
  H  T  U
  3  4  7
+ 2  8  4
---------
```

17
```
  H  T  U

+ _____
  ------
```

18
```
  H  T  U

+ _____
  ------
```

Complete the patterns.

19 Add 20.   8 28 [ ] [ ] [ ] [ ]

20 Make your own pattern by adding 9.

Answers

Addition p1

1 1 3 5 7 9 11 13 15 17 19

2 2 4 6 8 10 12 14 16 18 20

3 11　　　**4** 19　　　**5** 4　　　**6** 14

7 15 + 1　13 +3　11 + 5　9 + 7

8 14 + 2　12 + 4　10 + 6　(8 + 8)

9 −　**10** Any of these: 13 + 2, 11 + 4, 9 + 6, 7 + 8, 5 + 10, 3 + 12, 1 + 14

11 441　　**12** 442　　**13** 445

14 443　　**15** 444　　**16** 441　443　445

Harder addition p2

1 odd　　　**2** even　　　**3** even

4 even　　　**5** odd　　　**6** odd

7
```
  2 1      or      2 3
+ 4 3            + 4 1
-----            -----
  6 4              6 4  even
```

8
```
  1 2      or      3 2
+ 3 4            + 1 4
-----            -----
  4 6              4 6  even
```

9 417　　　**10** 305　　　**11** 439

12 434　　　**13** 426　　　**14** 427

15 405　　　**16** 408　　　**17** 459

18 6　　　　**19** 3

20 any addition with even answer

Subtraction p3

1 4　　　　**2** 3　　　　**3** 4

4 6　　　　**5** 7　　　　**6** 8

7 127

8 29　26　23　20　17　14　11　8　5　2

9 27　23　19　15　11　7　3

10 30　23　16　9　2

11 59　　　**12** 23　　　**13** 35　　　**14** 28

15 219　　**16** 138　　**17** 217

Harder subtraction p4

1 118　　**2** 114　　**3** 5　　　**4** 19

5 24　　　**6** 43　　　**7** 104　　**8** 32

9 22　　　**10** 12　　　**11** 44　　**12** 34

13 24　　　**14** 51　　　**15** 41　　**16** 31

17 217　　**18** 128　　**19** 237

20 254　　**21** 209　　**22** 411

Multiplication p5

1 6　12　18

2 6　12　18　24　30　36　42　48　54　60 coloured

3 24　　　**4** 36　　　**5** 54　　　**6** 42

7 48　　　**8** 60　　　**9** 30　　　**10** 18

11 132　　**12** 150　　**13** 138　　**14** 186

15 9 + 9 = 18　　　　**16** 9 + 9 + 9 = 27

　　2 × 9 = 18　　　　　　3 × 9 = 27

17 9 + 9 + 9 + 9 = 36

　　4 × 9 = 36

18 9 + 9 + 9 + 9 + 9 = 45

　　5 × 9 = 45

19 189　　**20** 288　　**21** 387　　**22** 225

Harder multiplication p6

1 6　12　18　24　30　36　42　48　54　60

2 9　18　27　36　45　54　63　72　81　90

3 18　36　54

4 6 × 9 = 54　　**5** 9 × 6 = 54

6 The answers are both the same.

7 6 × 37 = 222　**8** 6 × 29 = 174

9 9 × 25 = 225

10 126　　**11** 126　　**12** 234

13 270　　**14** 270　　**15** 477　　**16** 2

Division p7

1 12　　　　**2** 2

3 6　12　18　24　30　36　42　48　54　60

4 9　18　27　36　45　54　63　72　81　90

5 5　　　　**6** 9　　　　**7** 6

8 9　　　　**9** 4　　　　**10** 8

11 4　　　**12** 10　　　**13** 3　　　**14** 5

15 13　　　**16** 21　　　**17** 12　　　**18** 22

Harder

1 18 ÷ 2 = 9　9 × 2 = 18

　　18 ÷ 9 = 2　2 × 9 = 18

2 4　　　**3** 6

4 16　　　**5** 19　　　**6** 28　　　**7** 14

8 15　　　**9** 18　　　**10** 13　　　**11** 16

Money p8

1 £2·74　　**2** £5·63　　**3** £4·70

4 £5·25　　**5** £7·91　　**6** £7·53

7 25　　　**8** 8　　　**9** 6　　　**10** 3

11 Many answers are possible, e.g. £1, 50p, 20p, 2p

12 8p　　　**13** 25p　　　**14** 31p

Harder money p9

1 £4·84　　**2** £4·82　　**3** £5·73　　**4** large

5 10p, 5p　**6** £0·25　　**7** £1·04　　**8** £0·38

9 Many answers are possible.

Weight p10

1 153 g　　**2** 209 g　　**3** 245 g

4 695 g　　**5** 663 g　　**6** 615 g　　**7** 717 g

8 328 g　　**9** 209 g　　**10** 328 g　　**11** 329 g

Harder

1 680 g　　**2** 973 g　　**3** 625 g　　**4** Jo

5 Viv　　　**6** Tom　　　**7** 348 g　　**8** 55 g

Time p11

1 − **4** Clock faces showing 10:05, 3:30, 7:15, 4:25

5 Arrows to 5:15, 8:20, 11:00, 7:30

6 20 to 4　**7** 5 to 1　**8** 25 to 6　**9** 10 to 3

10 − **13** Clockfaces showing 8:40, 4:45, 11:25, 8:50

Harder time p12

1 – **7** to include clock faces

1 9:00, 9 o'clock **2** 5:35, 25 to 6

3 3:35, 25 to 4 **4** 12:10, 10 past 12

5 3:45, quarter to 4 **6** 10:55, 5 to 11

7 7:55, 5 to 8

Addition p13

1 381 **2** 781 **3** 838 **4** 845

5 421 **6** 911 **7** 730 **8** 824

9 621 **10** 531 **11** 751

12 100 + 30 + 2 = 132

13 300 + 10 + 4 = 314

14 400 + 20 + 1 = 421

15 4 9 14 19 24 29 34 39

16 9 19 29 39 49 59 69 79

Harder addition p14

1 200 40 3 → 243

2 400 30 4 → 434

3 500 60 7 → 567

4 600 50 3 → 653

5 342 **6** 421 **7** 635

8 809 **9** 700 **10** 820 **11** 902

12 910 **13** 551 **14** 710 **15** 311

16 631 **17** 801 **18** 930

19 8 28 48 68 88 108

20 pattern of 'add 9'

Subtraction p15

1 179	**2** 145	**3** 136	**4** 2
169	140	133	
159	135	130	
149	130	127	
139	125	124	
129	120	121	
119	115	118	

5 383 **6** 142 **7** 292 **8** 261

9 282 **10** 171 **11** 290 **12** 493

13 391 **14** 192 **15** 382

16 234 → 225 → 216 → 207 → 198

17 310 → 301 → 292 → 283 → 274

18 200 → 180 → 160 → 140 → 120

Harder subtraction p16

1 292 **2** 491 **3** 361 **4** 190

5 164 **6** 697 **7** 285 **8** 290

9 153 **10** 472 **11** 571

12 245 225 205 185 165 145

180 160 140 120 100 80 60

13 927 827 727 627 527

805 705 605 505 405

Multiplication p17

1 28 63 49 35

2 24 40 72 32 64 48

3 50 80 40

4 200 **5** 296 **6** 432 **7** 232

8 328 **9** 208 **10** 91 **11** 385

12 294 **13** 259 **14** 203 **15** 308

20 3 **21** 10 **22** 9

Harder multiplication p18

1 182 **2** 408 **3** 456 **4** 441

5 280 **6** 238 **7** 203 **8** 680

9 576 **10** 539 **11** 704 **12** 448

13 252 **14** 10 × number in class

15 184 **16** 126

17 8 **18** 7 **19** 2

20 5 **21** 8 **22** 7

23 8 **24** 8 **25** 7

26 7 **27** 10 **28** 6

29 8 **30** 6 **31** 8

32 4 **33** 3 **34** 8

Division p19

1 2 **2** 4 **3** 1

4 8 **5** 6 **6** 7 **7** 10

8 13 r 1 **9** 11 r 1 **10** 11 r 1 **11** 12 r 2

12 15 r 2 **13** 13 r 2 **14** 17 **15** 18 r 2

16 7 each, 1 left over

Harder

1 21 **2** 35 **3** 72

4 48 **5** 8 **6** 7

7 3 teams, 6 left over

8 3 teams, 3 left over

9 8 bags, 1 left over

10 2 r 5 **11** 6 r 3 **12** 7 r 4 **13** 6 r 1

Money p20

1 4 × 7p = 28p = £0·28

2 7 × 7p = 49p = £0·49

3 9 × 7p = 63p = £0·63

4 3 × 7p = 21p = £0·21

5 15p + 15p + 15p + 15p = £0·60

or 4 × 15p = £0·60

6 20p + 20p + 20p + 20p + 20p = £1·00

or 5 × 20p = £1·00

Harder

1 Total cost = £1·95

2 Total cost = £1·02

Change from £1·10 = £0·08

3 Total cost = £1·17

Change from £1·20 = £0·03

4 2

5 3

Addition p21

1

1000	10 hundreds
1800	18 hundreds
1100	11 hundreds
1500	15 hundreds
1900	19 hundreds
1700	17 hundreds

2 There are 8 possible pairs, e.g. 3973, 1684

3 1300 1600 1900 2100 2500 2900

4 2000 1800 1600 1400 1200 1000
5 1000 1100 1200 1300 1400 1500 1600 1700 1800
6 3681 **7** 7718 **8** 2909 **9** 8922
10 3631 **11** 5891 **12** 7546 **13** 7623
14 5357 **15** 1210 **16** 1411 **17** 3121

Harder addition p22

1

2 5952   **3** 7380
4 6500 5500 4500 3500 2500 1500
5 1100 2100 3100 4100 5100 6100 7100
6 The odd one out is 9723
   The others are all 9783
7 205 m   **8** 2211   **9** 5202   **10** 3042

Subtraction p23

1 236   **2** 206   **3** 363   **4** 595
5 279   **6** 256   **7** 243   **8** 279
9 264 → 1264 → 1364 → 1463
   156 → 1156 → 1256 → 1355
   300 → 1300 → 1400 → 1499
10 4348 → 3348 → 3248 → 3149
   5000 → 4000 → 3900 → 3801
   7311 → 6311 → 6211 → 6112
11–**13** There are 12 possible numbers, e.g.
   3694 → 4000
   6394 → 6000
   6934 → 7000
14 16   13   10   7   4   1 −2 −5 −8

Harder subtraction p24

1 189   **2** 199
3 179   **4** 529   **5** 285   **6** 283
7 2150 2250 2350 2450 2550 2650 2750
8 1825 1625 1425 1225 1025 825 625
9 open
10 7244   **11** 4527   **12** 6830   **13** 8535
14 2412   **15** 1615   **16** 7111   **17** 5768
18 3419 and 2617
19 2499

Multiplication p25

1 $2 \times 7 = 14$ or $7 \times 2 = 14$
2 $3 \times 7 = 21$ or $7 \times 3 = 21$
3 $6 \times 7 = 42$ or $7 \times 6 = 42$
4 $3 \times 9 = 27$ or $9 \times 3 = 27$
5 $4 \times 8 = 32$ or $8 \times 4 = 32$
6 $6 \times 8 = 48$ or $8 \times 6 = 48$
7 5   **8** 7   **9** 5
10 9   **11** 7   **12** 9
13 80   **14** 104   **15** 111   **16** 98
17 108   **18** 266   **19** 368

Harder

1 40   **2** 24   **3** 30
4 Some of 30, 35, 40, 42, 48, 56, 210, 240, 280, 336, 1680.
5 342   **6** 441   **7** 384   **8** 333

Division p26

1 131   **2** 142
3 128   **4** 138   **5** 226   **6** 338
7 171   **8** 283   **9** 351   **10** 464
11 120
12 221   **13** 321   **14** 161   **15** 182

Harder

1 133   **2** 112
3 183   **4** 149   **5** 124   **6** 163
7

¹1	9	²1
4		5
³2	6	9

Length p27

1

236 cm	2·36 m
69 cm	0·69 m
67 cm	0·67 m
419 cm	4·19 m
509 cm	5·09 m

2

1·74 m	174 cm
4·01 m	401 cm
3·86 m	386 cm
0·35 m	35 cm
2·40 m	240 cm

3 7·81   **4** 5·93   **5** 9·55   **6** 10·48
7 2·34   **8** 3·34   **9** 4·04   **10** 0·80
11 7·63 m   **12** 3·46 m

Harder

1 5·68 m + 2·92 m = 8·60 m
2 4·09 m − 2·92 m = 1·17 m
3

0·10 m	10 cm
0·01 m	1 cm
0·11 m	11 cm

4

7 cm	0·07 m
70 cm	0·70 m
700 cm	7·00 m

Weight p28

1

2·460 kg	2460 g
1·600 kg	1600 g
3·195 kg	3195 g
5·000 kg	5000 g
4·250 kg	4250 g

2

2750 g	2·750 kg
1340 g	1·340 kg
3700 g	3·700 kg
2000 g	2·000 kg
1275 g	1·275 kg

3 5·870   **4** 8·000   **5** 6·693
6 3·241 kg   **7** 6·322 kg

Harder

1 kg
```
  0 · 2 2 0
  2 · 1 6 0
+ 0 · 5 2 6
───────────
  2 · 9 0 6
```

2 kg
```
  0 · 1 0 5
  1 · 3 0 6
+ 0 · 1 5 0
───────────
  1 · 5 6 1
```

3 2·521 kg   **4** 1·220 kg   **5** 3·741 kg

Mental test 1 (covering pages 1–6)

1 Write two odd numbers that add up to 10 (1, 9; 3, 7; 5, 5)

2 Use 2, 3 and 4. Make the smallest 3-digit odd number (243)

Add 10 to each number: **3** 122 (132) **4** 107 (117) Add 20 to each number: **5** 142 (162) **6** 202 (222)

Find the difference between these numbers: **7** 6 and 10 (4) **8** 24 and 31 (7)

Subtract 9 from these numbers: **9** 85 (76) **10** 64 (55) Take 19 from these: **11** 76 (57) **12** 95 (76)

13 Four 9s (36) **14** Six 9s (54) **15** Eight 9s (72) **16** 6 + 6 + 6 (18) **17** 6 times 6 (36) **18** 7 lots of 6 (42)

Mental test 2 (covering pages 7–10)

1 How many groups of 6 are in 18 ? (3) **2** How many sets of 9 are in 36 ? (4) **3** Divide 18 by 9 (2)

4 Share 24 by 6 (4) **5** How many 2s in 22 ? (11) **6** 20 children get in groups of 4. How many groups? (5)

7 Add £1.50, 15p and 5p (£1.70) **8** Add £1.50 and £1.20 (£2.70) **9** How many 20p coins make £1.40? (7)

10 How many 50p coins make £2? (4) **11** I have £1.30. I spend 20p. How much is left? (£1.10)

12 I have £3. I spend £1.50. How much is left? (£1.50) **13** Add 100g and 30g (130g)

14 Add 100g, 200g and 40g (340g) **15** 400g + 37g (437g) **16** 300g take away 50g (250g)

17 240g minus 210g (30g) **18** How much heavier is 327g than 320g ? (7g)

Mental test 3 (covering pages 11–16)

Write digital times: **1** Twenty past 4 (4:20) **2** 9 o'clock (9:00) **3** 10 to 2 (1:50)

Write o'clock times: **4** 4:05 (5 past 4) **5** 11:45 (quarter to 12) **6** 8:30 (half past 8)

Write the next number: **7** 2, 6, 10, ... (14) **8** 22, 42, 62, ... (82) **9** Add 200, 5 and 60 (265)

10 What is the number 1 worth in 217 ? (10) **11** 29 plus 4 (33) **12** 38 and 6 (44)

What is next? **13** 91, 71, 51, ... (31) **14** 735, 635, 535, ... (435) **15** 180 minus 100, take away 10 (70)

16 The difference between 24 and 16 (8) **17** What is the pattern? 226, 216, 206, 196 (minus 10)

18 There are 32 children in a class. 7 are away. How many are at school? (25)

Mental test 4 (covering pages 17–20)

1 6 multiplied by 7 (42) **2** 8 times 7 (56) **3** 2 lots of 8 (16) **4** 9 multiplied by 8 (72)

5 12 multiplied by 10 (120) **6** 5 lots of 8 (40) **7** How many 7s in 42? (6) **8** 70 divided by 7 (10)

9 Divide 16 by 3. What is the remainder? (1) **10** How many 8s in 64? (8) **11** Divide 40 by 8 (5)

12 Share 24 sweets between 8 children (3) Write as pounds: **13** 150p (£1.50) **14** 105p (£1.05)

Find the answer in pounds: **15** Add 22p, 22p and 22p (£0.66) **16** 4 items at 8p each (£0.32)

17 9 items at 9p (£0.81) **18** I have £1.50. I buy 3 tickets at 30p each. How much is left? (£0.60 or 60p)

Mental test 5 (covering pages 21–25)

1 How many hundreds in 1000? (10) **2** How many hundreds in one thousand eight hundred? (18)

3 Write four thousand and eighty-six (4086) **4** 500 + 800 (1300)

Write the next number: **5** 1200, 1400, 1600, ... (1800) **6** 4500, 3500, 2500, ... (1500)

7 Add 100 to 157 (257) **8** Add 99 to 345 (444) **9** Subtract 1000 from 3245 (2245)

10 Take 99 from 105 (6) Round to the nearest 1000: **11** 2800 (3000) **12** 4129 (4000)

13 3 times 8 (24) **14** 9 times what is 27? (3) **15** Six 9s (54) **16** Seven 4s (28)

17 The factors of 5 are 1 and what ? (5) **18** What are the four factors of 10 ? (1, 10, 2, 5)

Mental test 6 (covering pages 26–28)

1 35 divided by 5 (7) **2** 32 divided by 4 (8) **3** What divided by 4 gives 7 ? (28)

4 49 divided by what gives 7 ? (7) **5** There are 40 children. How many rows of 8 ? (5)

6 32 balls are put into boxes of 6. How many are left over? (2) **7** How many cm in 3 m? (300)

8 How many cm in one and a half metres? (150)

Write as metres: **9** 13 cm (0.13 m) **10** 3 cm (0.03 m) **11** 110 cm (1.10 m) **12** 205 cm (2.05 m)

13 How many g in 4 kg? (4000) **14** A parcel weighs two and a half kg. How many g is that ? (2500)

15 Write as kg 1200 g (1.200 kg) **16** What is the total weight of 1 kg and 400 g? (1.400 kg or 1400 g)

17 Write 2050 g in kg (2.050 kg) **18** What is the difference between 1 kg and 750 g? (250 g or 0.250 kg)

Subtraction (HTU, patterns)

Complete the patterns.

1 Subtract 10

179
169

2 Subtract 5

145

3 Subtract 3

136

4 Subtract ☐

130
128
126
124
122
120
118

5
```
  H  T  U
  5  5  5
- 1  7  2
---------
```

6
```
  H  T  U
  3  2  4
- 1  8  2
---------
```

7
```
  H  T  U
  4  8  3
- 1  9  1
---------
```

8
```
  H  T  U
  6  3  6
- 3  7  5
---------
```

9
```
  H  T  U
  8  4  5
- 5  6  3
---------
```

10
```
  H  T  U
  7  6  7
- 5  9  6
---------
```

11
```
  H  T  U
  5  8  2
- 2  9  2
---------
```

12
```
  H  T  U
  9  3  7
- 4  4  4
---------
```

13 573 − 182

14 343 − 151

15 778 − 396

16 Subtract 9.   234 → ☐ → ☐ → ☐ → ☐

17 Subtract 9.   310 → ☐ → ☐ → ☐ → ☐

18 Subtract 20.   200 → ☐ → ☐ → ☐ → ☐

Harder subtraction

1
```
H  T  U
5  8  3
- 2  9  1
---------
```

2
```
H  T  U
6  2  1
- 1  3  0
---------
```

3
```
H  T  U
7  3  5
- 3  7  4
---------
```

4
```
H  T  U
4  7  0
- 2  8  0
---------
```

5
```
H  T  U
4  5  6
- 2  9  2
---------
```

6
```
H  T  U
8  8  8
- 1  9  1
---------
```

7
```
H  T  U
6  4  8
- 3  6  3
---------
```

8
```
H  T  U
5  1  1
- 2  2  1
---------
```

9 435 − 282

10 627 − 155

11 816 − 245

12 Subtract 20 in each pattern.

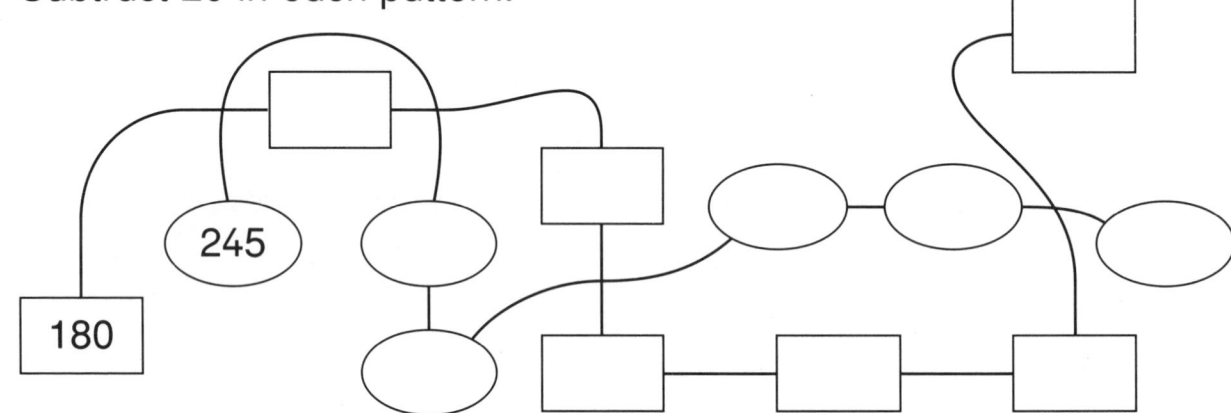

13 Subtract 100 in each pattern.

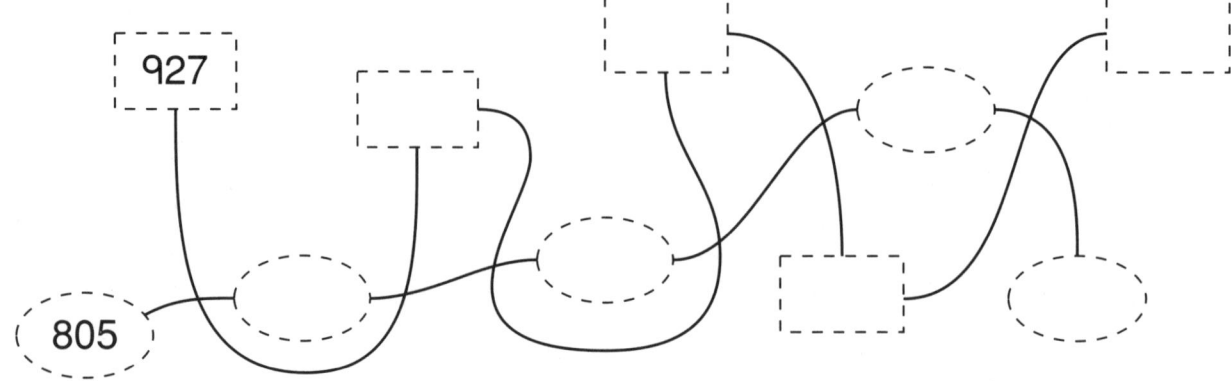

Multiplication (by 7, 8)

Complete the table roads.

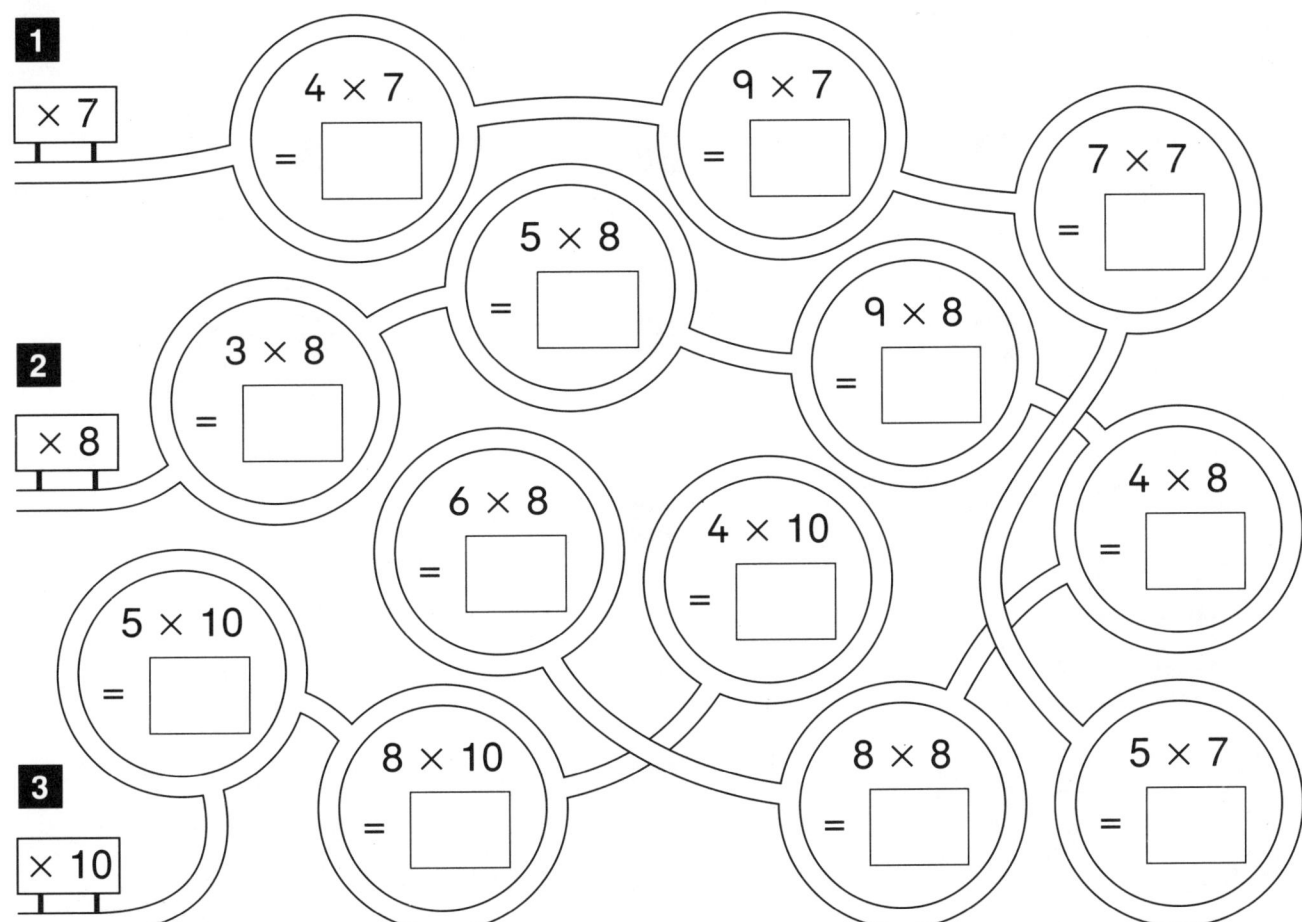

4
   2 5
×   8
_____

5
   3 7
×   8
_____

6
   5 4
×   8
_____

7
   2 9
×   8
_____

8
   4 1
×   8
_____

9
   2 6
×   8
_____

10
   1 3
×   7
_____

11
   5 5
×   7
_____

12
   4 2
×   7
_____

13
   3 7
×   7
_____

14
   2 9
×   7
_____

15
   4 4
×   7
_____

16 ☐ × 10 = 30   **17** ☐ × 6 = 60   **18** ☐ × 10 = 90

Harder multiplication

1
```
    2 6
  ×   7
```

2
```
    5 1
  ×   8
```

3
```
    5 7
  ×   8
```

4
```
    6 3
  ×   7
```

5
```
    3 5
  ×   8
```

6
```
    3 4
  ×   7
```

7
```
    2 9
  ×   7
```

8
```
    8 5
  ×   8
```

9
```
    7 2
  ×   8
```

10
```
    7 7
  ×   7
```

11
```
    8 8
  ×   8
```

12
```
    6 4
  ×   7
```

13 How many days are there in 36 weeks ?

14 How many toes are there in your class ?

15 How many legs do 23 spiders have ?

16 There are 7 players in a netball team. How many players in 18 teams ?

17 $3 \times \boxed{} = 24$

18 $5 \times \boxed{} = 35$

19 $\boxed{} \times 7 = 14$

20 $\boxed{} \times 10 = 50$

21 $5 \times \boxed{} = 40$

22 $7 \times \boxed{} = 49$

23 $\boxed{} \times 10 = 80$

24 $6 \times \boxed{} = 48$

25 $\boxed{} \times 4 = 28$

26 $9 \times \boxed{} = 63$

27 $9 \times \boxed{} = 90$

28 $\boxed{} \times 7 = 42$

29 $4 \times \boxed{} = 32$

30 $\boxed{} \times 10 = 60$

31 $9 \times \boxed{} = 72$

32 $\boxed{} \times 10 = 40$

33 $7 \times \boxed{} = 21$

34 $\boxed{} \times 3 = 24$

Division (by 7, 8, remainders)

Choose the correct answer.

1 14 ÷ 7 = ☐   **2** 28 ÷ 7 = ☐

3   7 ÷ 7 = ☐   **4** 56 ÷ 7 = ☐

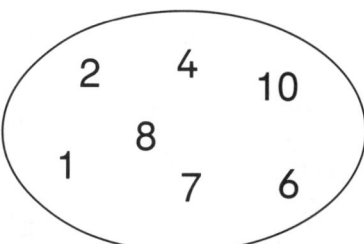

5 48 ÷ 8 = ☐   **6** 56 ÷ 8 = ☐   **7** 80 ÷ 8 = ☐

8 2)27   **9** 2)23   **10** 3)34   **11** 3)38

12 3)47   **13** 3)41   **14** 3)51   **15** 3)56

16 Share 22 apples between 3 people.

How many each do they get? ☐   How many are left over? ☐

Harder

1 ☐ ÷ 7 = 3   **2** ☐ ÷ 7 = 5   **3** ☐ ÷ 8 = 9

4 ☐ ÷ 8 = 6   **5** 56 ÷ ☐ = 7   **6** 49 ÷ ☐ = 7

There are 27 children.

7 How many teams of 7? ☐   **9** There are 65 bananas.

How many left over? ☐   How many bags of 8? ☐

8 How many teams of 8? ☐   How many are left over? ☐

How many left over? ☐

10 19 ÷ 7 = ☐ r ☐   **11** 45 ÷ 7 = ☐ r ☐

12 60 ÷ 8 = ☐ r ☐   **13** 49 ÷ 8 = ☐ r ☐

Money (multiplication of pence)

School Fair

stickers
7p each

rulers
20p each

pencils 15p each

Find the cost of

1 4 stickers. 4 × 7p = ☐ p = £ ☐ · ☐

2 7 stickers. ☐ × 7p = ☐ p = £ ☐ · ☐

3 9 stickers. ☐ × ☐ p = ☐ p = £ ☐ · ☐

4 3 stickers. ☐ × ☐ p = ☐ p = £ ☐ · ☐

Find the cost of these in two ways

5 4 pencils. ☐ p + ☐ p + ☐ p + ☐ p = £ ☐ · ☐

or 4 × ☐ p = £ ☐ · ☐

6 5 rulers ☐ p + ☐ p + ☐ p + ☐ p + ☐ p = £ ☐ · ☐

or ☐ × ☐ p = £ ☐ · ☐

Harder

Fair Tickets

45p	adult
30p	child
12p	programme

Find the total cost of

1 ☐ adult ☐ adult ☐ adult ☐ child ☐ child

Total cost = £ ☐ · ☐

2 ☐ child ☐ child ☐ child ☐ programme

Total cost = £ ☐ · ☐

Change from £1·10 = £ ☐ · ☐

3 ☐ adult ☐ child ☐ child ☐ programme

Total cost = £ ☐ · ☐

Change from £1·20 = £ ☐ · ☐

4 How many adult tickets can be bought for £1? ☐

5 How many children's tickets can be bought for £1? ☐

Addition (ThHTU, place value, patterns)

1 Fill in the missing numbers.

1000	10 hundreds
	18 hundreds
	11 hundreds
1500	
1900	
	17 hundreds

2

(3000) (900) (3) (600)
(70) (80) (1000) (4)

Use each of these numbers to make two thousands numbers.

3 Write these numbers in order. Start with the smallest.

1900   1600   2500   2900   2100   1300

Finish these number patterns.

4 2000   1800   1600   [ ]   [ ]   [ ]

5 1000   1100   [ ]   [ ]   [ ]   [ ]   [ ]   1700   1800

6
```
  2 4 6 5
+ 1 2 1 6
---------
```

7
```
  3 5 7 3
+ 4 1 4 5
---------
```

8
```
  1 6 2 9
+ 1 2 8 0
---------
```

9
```
  3 7 5 6
+ 5 1 6 6
---------
```

10
```
  1 3 5 3
+ 2 2 7 8
---------
```

11
```
  2 4 4 7
+ 3 4 4 4
---------
```

12
```
  6 1 7 4
+ 1 3 7 2
---------
```

13
```
  4 3 7 8
+ 3 2 4 5
---------
```

14
```
  6 8 7 9
- 1 5 2 2
---------
```

15
```
  8 6 3 4
- 7 4 2 4
---------
```

16
```
  5 9 4 1
- 4 5 3 0
---------
```

17
```
  4 3 3 5
- 1 2 1 4
---------
```

Harder addition (and subtraction)

1 Draw arrows to match the numbers in the boxes.

200
2
20
2000

9256	6529
5692	2965

9000
900
90
9

Write the following in numbers.

2 Five thousand nine hundred and fifty two ☐

3 Seven thousand three hundred and eighty ☐

Complete the patterns.

4 6500 5500 4500 ☐ ☐ ☐

5 1100 2100 3100 ☐ ☐ ☐ ☐

6 Find the totals. Ring the odd one out.

```
   6 7 6 5          5 5 9 2          7 0 8 6
 + 3 0 1 8        + 4 1 9 1        + 2 6 9 7
 _____        _____        _____


   8 3 5 9          9 2 8 3          4 3 6 8
 + 1 4 2 4        +   5 0 0        + 5 3 5 5
 _____        _____        _____
```

7 A mountain is 1987 m high.
Rebecca climbs 1782 m.
How far is she from the top?

```
 8  3 4 2 8       9  7 6 1 3      10  5 0 5 6
  - 1 2 1 7        - 2 4 1 1        - 2 0 1 4
  _____        _____        _____
```

Subtraction (ThHTU, rounding, patterns)

1
```
   3  6  4
 - 1  2  8
 ─────────
```

2
```
   4  5  1
 - 2  4  5
 ─────────
```

3
```
   9  3  5
 - 5  7  2
 ─────────
```

4
```
   7  1  6
 - 1  2  1
 ─────────
```

5
```
   5  3  7
 - 2  5  8
 ─────────
```

6
```
   4  4  4
 - 1  8  8
 ─────────
```

7
```
   6  4  1
 - 3  9  8
 ─────────
```

8
```
   4  1  2
 - 1  3  3
 ─────────
```

Complete the tables.

9

Number	Add 1000	Add 100	Add 99
264 ⟶	1264 ⟶	1364 ⟶	
156			
300			

10

Number	Subtract 1000	Subtract 100	Subtract 99
4348 ⟶	3348 ⟶	⟶	⟶
5000			
7311			

Arrange the numbers to make three even numbers.
Round each answer to the nearest 1000.

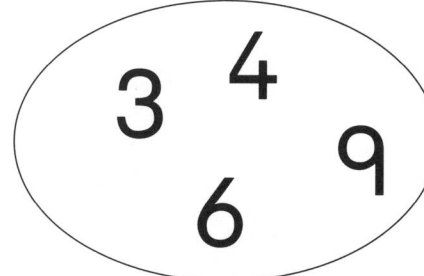

11

12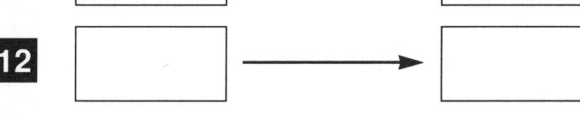

13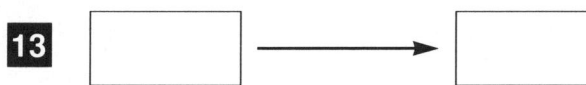

14 Complete the pattern.

16   13   10   7   ▢ ▢ ▢ ▢ ▢

Harder subtraction

1 There are 365 children in a school.
176 are girls.
How many are boys?

2 166 of the children in the school are infants.
How many are juniors?

3
```
  3 4 5
- 1 6 6
```

4
```
  7 0 8
- 1 7 9
```

5
```
  6 4 2
- 3 5 7
```

6
```
  8 4 1
- 5 5 8
```

Finish the patterns.

7 2150 2250 ☐ ☐ ☐ ☐ 2750

8 1825 1625 ☐ ☐ ☐ ☐ 625

9 Write your own pattern. Start at 2000. Finish at 1000.

10 Add 1000

6244 ⟶ ☐

11 Add 999

3528 ⟶ ☐

12 Add 999

5831 ⟶ ☐

13 Add 999

7536 ⟶ ☐

14 Subtract 1000

3412 ⟶ ☐

15 Subtract 999

2614 ⟶ ☐

16 Subtract 999

8110 ⟶ ☐

17 Subtract 999

6767 ⟶ ☐

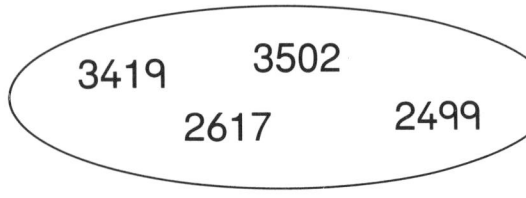

3419   3502
    2617   2499

18 Which of the numbers round to 3000?

19 Which of the numbers round to 2000?

Multiplication (tables to 10, factors)

You need a multiplication square.
Find the factors.

1 ☐ × ☐ = 14

2 ☐ × ☐ = 21

3 ☐ × ☐ = 42

4 ☐ × ☐ = 27

5 ☐ × ☐ = 32

6 ☐ × ☐ = 48

Find the missing factor.

7 ☐ × 10 = 50

8 ☐ × 7 = 49

9 8 × ☐ = 40

10 ☐ × 8 = 72

11 8 × ☐ = 56

12 ☐ × 1 = 9

13
```
    1  6
×      5
_____
```

14
```
    2  6
×      4
_____
```

15
```
    3  7
×      3
_____
```

16
```
    4  9
×      2
_____
```

17 27 × 4 = ☐

18 38 × 7 = ☐

19 46 × 8 = ☐

Harder

Find the numbers that have all the factors.

1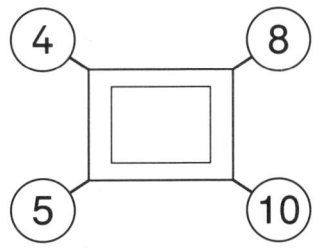

2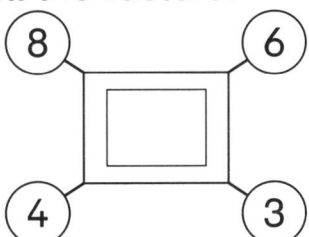

3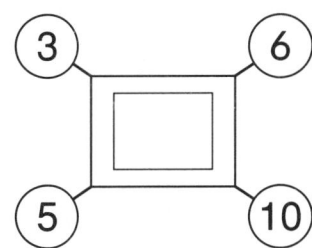

4 Make numbers using 2 or more of these factors.

5   6   7   8

5
```
    5  7
×      6
_____
```

6
```
    6  3
×      7
_____
```

7
```
    4  8
×      8
_____
```

8
```
    3  7
×      9
_____
```

Division (HTU by 1 digit)

1 A worker makes 262 socks.
How many pairs of socks? $2\overline{)262}$

2 Another worker makes 284 socks.
How many pairs is this?

3 $2\overline{)256}$ **4** $2\overline{)276}$ **5** $2\overline{)452}$ **6** $2\overline{)676}$

7 $2\overline{)342}$ **8** $2\overline{)566}$ **9** $2\overline{)702}$ **10** $2\overline{)928}$

11 A factory makes 360 juggling balls.
How many sets of 3 are there?

12 $3\overline{)663}$ **13** $3\overline{)963}$ **14** $3\overline{)483}$ **15** $3\overline{)546}$

Harder

1 A farmer grows 532 peaches.

How many packets of 4?

2 A baker bakes 672 cakes.

How many trays of 6?

3 $4\overline{)732}$ **4** $5\overline{)745}$ **5** $7\overline{)868}$ **6** $6\overline{)978}$

7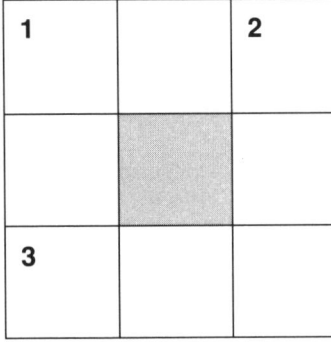

Across
1 $764 \div 4$
3 Eight hundred and seven shared by three

Down
1 Divide eight hundred and fifty two by six
2 $795 \div 5$

Length (decimal notation, addition/subtraction)

Complete the charts.

1

236 cm	2·36 m
69 cm	
67 cm	
419 cm	
509 cm	

2

1·74 m	174 cm
4·01 m	
3·86 m	
0·35 m	
2·40 m	

3
```
    m
  5·37
+ 2·44
_____
```

4
```
    m
  4·85
+ 1·08
_____
```

5
```
    m
  3·90
+ 5·65
_____
```

6
```
    m
  2·76
+ 7·72
_____
```

7
```
    m
  6·29
- 3·95
_____
```

8
```
    m
  5·80
- 2·46
_____
```

9
```
    m
  7·61
- 3·57
_____
```

10
```
    m
  9·42
- 8·62
_____
```

11 Add 6·08 m

to 1·55 m

12 Subtract 1·37 m

from 4·83 m

Harder

1 Use the lengths.
Which two add
up to 8·60 m?

```
    m

+ _____
  _____
```

2 Which two lengths have
a difference of 1·17 m?

```
    m

- _____
  _____
```

5·68 m

4·09 m

2·92 m

Complete the charts.

3

0·10 m	cm
0·01 m	cm
0·11 m	cm

4

7 cm	m
70 cm	m
700 cm	m

Weight (decimal notation, addition/subtraction)

Complete the charts.

1

2·460 kg	2460 g
1·600 kg	
3·195 kg	
5·000 kg	
4·250 kg	

2

2750 g	2·750 kg
1340 g	
3700 g	
2000 g	
1275 g	

3 kg
```
   3 · 6 4 7
 + 2 · 2 2 3
 _____
```

4 kg
```
   4 · 2 1 0
 + 3 · 7 9 0
 _____
```

5 kg
```
   5 · 0 3 6
 + 1 · 6 5 7
 _____
```

Find the differences.

6   4·852 kg
      1·611 kg

7   7·648 kg
      1·326 kg

Harder

Add the weights.

1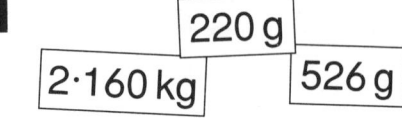

kg
```
    ·
    ·
 +  ·
 _____
    ·
```

2

kg
```
    ·
    ·
 +  ·
 _____
    ·
```

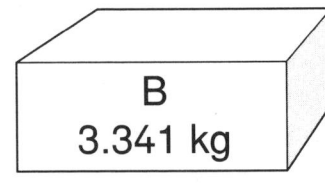

A
5.862 kg

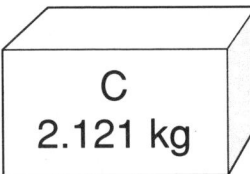

B
3.341 kg

C
2.121 kg

3 How much heavier is Box A than Box B?

4 How much lighter is Box C than Box B?

5 What is the difference between the heaviest and lightest box?